This book belongs to

_____,

a new friend of Rusty the Ranch Horse

Rusty is a Wyoming native. He was born near the beautiful mountains of Laramie. He has his own way of looking at life as a horse. He lives in the moment, wants to do the right thing, dances in the storms and is mostly interested in grazing. When you gain his trust he will be a friend for life.

Rusty Under the Western Skies
A Rusty the Ranch Horse Tale

© 2016 by Mary Fichtner
All rights reserved

Illustrations by Roz Fichtner

Layout and Design by Andy Grachuk
www.JingotheCat.com

Rusty Under the Western Skies
A Rusty the Ranch Horse Tale

Have you seen the grand show of a storm in the sky?

You can smell a storm too if you give it a try

That's how a horse knows when a billow is coming

Sensing it clearly like loud distant drumming

That's exactly what happened in the heat of that day

As Rusty grazed quietly the blue sky turned gray

The scent of the storm gave his sniffer a tickle

The way a soft feather can cause a slim prickle

It caused him to wonder if a smell could be wet

As it turned in his nose like a mini pirouette

The breeze brought fragrances in the soft wind

They floated like notes from a tuned violin

The air mostly hinted of grass freshly cut

Mixed in with aroma of hickory nut

Such smells are the kind that brought his nose pleasure

So he paused for a moment to enjoy the sweet treasure

Then a twisting gust stirred up his mane and his tail

The tingle it caused down his back left a trail

The skinny fine hairs on his legs felt a prick

Like a cold spinning wisp that puts out a small wick

The clouds crazy colors began to unfold

They looked like mixed paint, a sight to behold

The sky filled with hues of black, blue and steel

As the storm began rolling like a big wagon wheel

A new fragrance drifted into Rusty's muzzle

It resembled sweet tea that you just want to guzzle

A puff of the wind made him take a long whiff

A breeze mixed with scents make a horse want to sniff

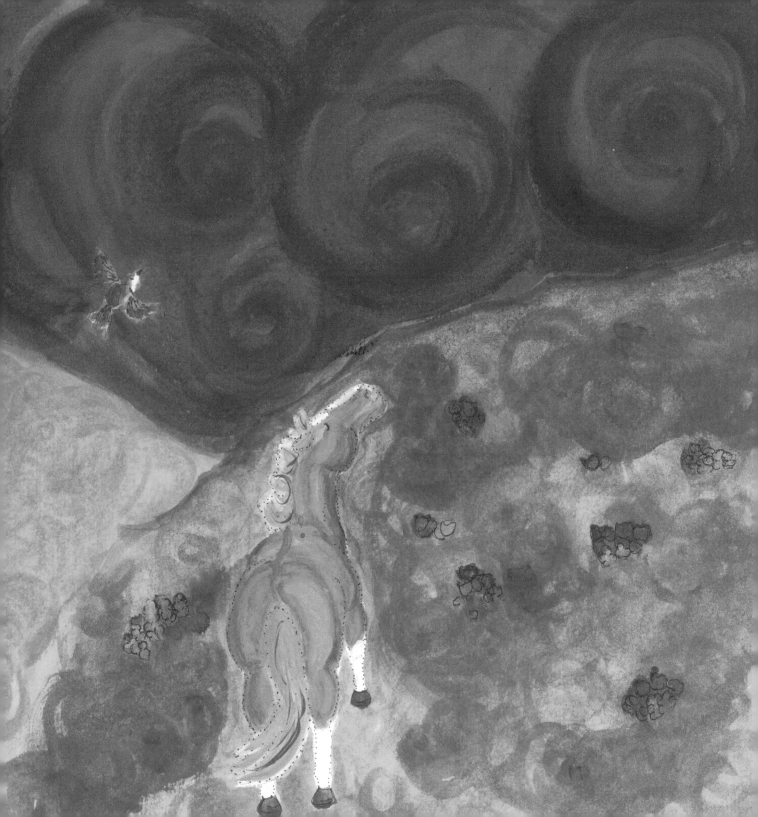

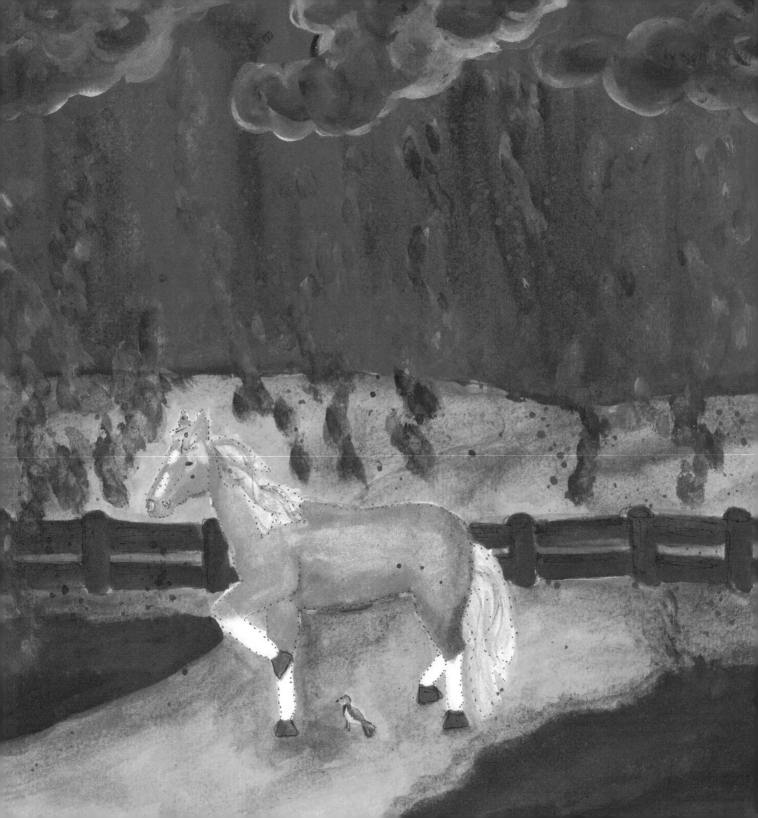

All the interesting smells caused Rusty to note

He was feeling cold air down in his throat

That's when the rain started hammering down

And you'd think on his face there would be big frown

It pelted and soaked his shiny orange hair

But he didn't much act like he gave it a care

He noticed the barn was open and dry

He had a quick thought to give it a try

But it left just as fast as the storm had rolled in

As he felt the bold force of the cold crazy wind

He decided to race straight into that gale

His pace caused the rush to whip up his blonde tail

Then Rusty jumped up with his nose to the ground

Cut the air right in two with a thud for the sound

It happened just when a lightning bolt struck

And that funny horse thought the sound came from his buck

The rain poured down hard upon his strong back

The thunder was loud like a whip that you crack

Rusty danced in the squall, the earth his big stage

If it was a show it would be all the rage

The strength that he used as he danced in the storm

Made him want to show off crazy moves and perform

The prickles kept coming all over his skin

As the water flew off in the shape of a fin

His pasture was taking a new kind of shape

Right in the middle it looked like a lake

Then Rusty decided to slide to a stop

As the rain changed from heavy to barely a drop

He paused for a moment and took in the clear air

As he held his head high his nostrils did flare

All those smells that stirred up the inside of his nose

Quickly just softened like an opening rose

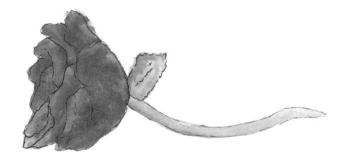

The crazy thing was when the storm moved on by

Everything became still without even a sigh

Though Rusty was soaked, his hair twisted in knots

He went right back to grazing with very few thoughts

Of the huge hurly burly that stirred up his day

And turned his fenced pasture into a ballet

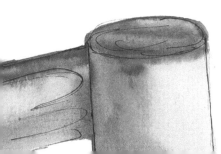

He didn't much care that the next one would too

As he pulled out more grass and gave it a chew

A rainbow was left as the big grand finale

It gracefully touched each side of the valley

The hot sun returned and dried off Rusty's hair

He's just a good horse without much of a care

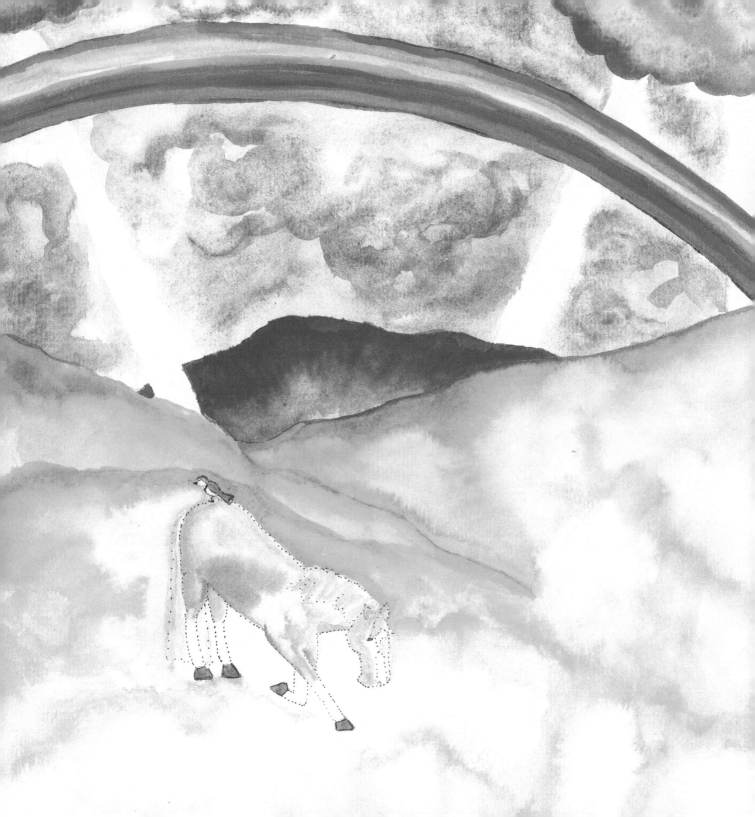

- Written by Mary Fichtner
- Illustrated by Rozlyn Fichtner
- In collaboration with Rusty the Ranch Horse

Mary Fichtner grew up with horses which gave her a bold western identity. Rodeo was her sport from 10 years old through college. She is a graduate of the University of Wyoming and her cowgirl spirit has grown stronger through life experiences and living in the cowboy state. The strong roots of the western lifestyle and horses gave her the character to manage the challenges and change required for a military wife and mother of four. Mary learned why horses have always owned her heart after being certified in equine assisted learning to help veterans and families. "They truly teach us about ourselves and others and heal broken hearts." To be able to share her love of horses and have her daughter illustrate her tales is the grandest adventure. Her hope is that you will laugh and smile as you see a little bit of life through a horse's eyes and learn to dance in your storms.

Rozlyn Fichtner grew up in a horse loving family in Wyoming and developed her own passion foreverything equine. She has competed in just about every horse event offered in the great cowboy state from jumping to showmanship to rodeo. She is attending college at the University of Wyoming where she competes on the rodeo team and is pursuing a degree in Art Education. Her talent in art and design along with her love of horses brought her to an unplanned adventure of illustrating a book about one of her favorites. Rozlyn and Rusty have shared many ribbons and adventures and through her skills with watercolors she hopes to share him with readers all over the world. "Everyone should meet Rusty!"

Other Rusty the Ranch Horse Tales

Rusty Goes to Frontier Days   2016

Rusty and His Saddle 2017

Rusty and the Pot of Gold 2017

Rusty and the River   2018

Wrong Color Rusty   2018

CPSIA information can be obtained
at www.ICGtesting.com
Printed in the USA
BVHW02n0300171018
530216BV00002B/17/P